MENTAL EXERCISE FOR DOGS

TABLE OF CONTENTS

INTRODUCTION

We all agree that a dog's health depends on having enough physical activity, but mental exercise is less frequently stressed. Dogs are incredibly intelligent and curious, and humans often underestimate their abilities.

As a result, dog owners should spend time playing activities with their dogs that promote mental abilities like focus, memory, and problem-solving.

Playing mental games with dogs can enhance their attitude, behavior, and sleep pattern.

Mental stimulation can help your dog feel exhausted and ensure a good night's sleep at the end of the day!

Just as kids tire out after a day of learning new skills at school or as adults might feel exhausted after a long week at work, so too can our furry companions!

Before you start reading, scan this QR Code to get all bonus content!

CHAPTER 1: EXERCISE BENEFITS FOR YOUR DOG'S PHYSICAL AND MENTAL HEALTH

Our dogs' physical and mental health must be maintained by regular exercise. It keeps them physically healthy, lessens anxiety and stress, and can lift their spirits.

In this section, We'll review five important advantages of regular exercise for your dog's physical and mental health.

5 Health Benefits of Exercise for Your Dog's Physical and Mental

Improved physical health is one of dogs' most evident advantages of regular exercise. Exercise can lower your dog's risk of injury and arthritis because it keeps your dog's muscles and joints healthy and flexible (especially for seniors). Additionally, exercise aids in keeping your dog at a healthy weight, lowering the risk of obesity and related health issues. The American Kennel Club states that regular exercise can improve your dog's cardiovascular health by boosting blood flow and oxygenation to the heart and lungs.

Mental Health Matters: Boosting Your Dog's Mood Through Exercise

Your dog can release any stored energy via physical activity, making them feel less stressed and anxious. Exercise also causes your dog's brain to produce endorphins, which enhance emotions of happiness and well-being. Gregory Berns, a neuroscientist at Emory University, conducted research in which he found that dogs can benefit from the same mental health advantages as humans do. Therefore, regular exercise can benefit your dog's mental health like it does us.

Improved Socialization: Regular Exercise Can Boost Your Dog's Interaction with Others

The socialization of your dog might also improve from regular exercise. They can interact with humans and other dogs by going on frequent walks, visiting the park, or attending doggie playdates. This can enhance their social abilities and lessen feelings of loneliness and isolation. Since well-socialized dogs are more likely to be self-assured and well-adjusted, socialization also aids in the prevention of behavioral problems like aggressiveness or fearfulness. Not to mention that taking your dog on walks or hikes with other dogs can be a wonderful opportunity for him to meet new people.

Enhanced Cognitive Function: Regular Exercise Can Maintain Your Dog's Mental Sharpness

Regular physical activity with your dog can also benefit their cognitive health. Dogs' minds require mental stimulation just like humans to stay sharp. Exercises that test their mental faculties, such as training or agility drills, can enhance their memory and problem-solving abilities.

Your dog's intellect can remain active and engaged by exposing them to different locations, sights, sounds, and scents when on walks or hikes. According to the American Kennel Club, regular exercise can even delay the development of age-related cognitive decline, such as cognitive dysfunction syndrome (CDS), and keep your dog's mind bright long into old life.

Good Behavior Starts with Exercise

Improved behavior is another advantage of regular exercise for dogs. Lack of activity can cause dogs to get restless or bored, resulting in destructive behavior like chewing, barking, and digging.

Regular exercise minimizes the incidence of these behaviors and keeps your dog cognitively active. If you must leave your dogs at home for more than a few hours, take your dog for a walk or a nice play session before you leave. This activity allows your dog to burn off energy and can help lessen their separation anxiety, which may be the underlying factor in behavioral problems.

Our dogs' mental and physical well-being must be maintained by regular exercise. Not only does it keep people physically active, but it also helps them feel better overall by lowering tension and anxiety.

Whether it's a lengthy walk, a swim in the lake, or a game of fetch, exercise in your dog's routine can enhance their mental and physical health and benefit them for many years.

CHAPTER 2: HOW TO UNDERSTAND YOUR DOG'S BODY LANGUAGE

Dogs communicate their moods and intentions via a wide range of peculiar body language clues. When compared to human communication, it may be extremely different.

It's important to comprehend what these sounds mean as whines, barks, and growls make up an important part of canine communication. On the other hand, dogs mostly depend on nonverbal clues from their body language. This may cause several dog-human miscommunications. People don't have tails, thus it may be sometimes difficult to understand dog body language. Sometimes it stands in sharp contrast to what a person would interpret the same signal, such as yawning or turning away. Learn some tricks to understanding canine body language to improve communication with your canine friend.

Tail-Wagging

The tail-wagging gesture seems to be a clear body language message. A dog is said to be happy if its tail is wagging, right? Wrong. People often misinterpret this indication.

A wagging tail indicates that the dog is feeling stimulated emotionally. Alternatively, it might be frustration or something even worse. The dog's tail position, pace, and direction of the wag can all be used to decipher the dog's intentions and emotions.

In general, the dog becomes more excited the faster the tail wags. When your dog greets you, it often wags its entire body while performing those long, slow, side-to-side tail sweeps. That is a relaxed dog. A higher level of alertness, maybe in a negative aspect, is indicated by a quicker, twitch-like wag. Think of a guard dog on alert.

Additionally, the wag's direction could provide hints. According to research, dogs prefer to wag their tails more to the right when feeling good about anything, such as engaging with their owner. Dogs' tails wag more to the left when they are being challenged. The dog then does the helicopter tail wag, in which the tail spins around. That is unquestionably a joyful wag. It often happens when a dog welcomes a special person.

The dog's tail posture in respect to the ground reveals important details about its mental state. Generally speaking, a dog is more assertive the higher its tail is. Dogs that tuck their tails between their knees or tip them downward toward the ground are showing signs of fear and anxiety. Dogs who are confident or even aggressive raise their tails up like a flag.

Dogs that are at ease maintain their tails in a neutral position. However, neutrality varies by breed. While certain breeds, like the Italian Greyhound, have a very low neutral tail posture, others, like the Chow Chow, have tails that naturally curve over their backs. If you become familiar with the neutral tail position, you can quickly spot a change in your dog's emotions.

Raised Hackles

The hair on a dog's back will rise when their hackles are aroused. The dog can puff up over the shoulders or down the back and to the tail, and it is technically called piloerection. This unmistakably indicates that the dog is excited, although perhaps not badly. Both excitement and strong curiosity are possible in addition to distress or tension in the dog. It often triggers an uncontrollable reaction in people, like goosebumps.

Posture

A dog's weight distribution can reveal a lot about its attitude and goals. Think of a dog that is crouched over and frightened. That indicates tension or dread. The dog seems smaller in this stance, which might indicate that it is trying to get away from something.

A dog rolling onto their back and showing their belly is the extreme of this stance. This might appear to be a dog asking for a belly rub, which is often the case with relaxed dogs. However, it could potentially be an indication of severe stress and worry. The dog could even pee a little to satisfy.

A dog adopts the opposite position standing with its weight pushed forward. This dog is approaching something. This might simply be the dog showing curiosity. However, combined with other aggressive body language indicators like a twitching tail held high it could also suggest hostile intentions. The dog is trying to look larger in this instance.

The play bow is a simple component of dog body language. Dogs do this when they lift their rump and place their chest on the ground. As the name suggests, it initiates play between dogs and humans.

The raising of the paw is a less clear indicator. The paw lift is a component of pointing behavior in pointing breeds like the English Setter when the dog points at nearby prey. However, when not in this context, a dog's lifted paw is usually indicative of doubt or even insecurity.

Facial Expressions

Although dog and human facial features have certain similarities, they are not used in the same ways. Dogs yawn for no apparent reason, but humans yawn when they feel either exhausted and bored. Dogs use yawning as a relaxing mechanism for themselves in stressful situations as well as to soothe other people, including their owners. She advises yawning at your dog to calm it during anxious situations, such as a visit to the vet. But do not be surprised if your dog yawns in response. Dogs can "catch" yawns just like people can.

Another dog body language cue that people sometimes misread is lip-licking. Dogs lick their lips after a wonderful meal, much like people do, but they also do it when nervous. It might be difficult to notice the tongue flick because it happens quickly. Your dog isn't expressing a desire to lick your face; your dog is expressing displeasure with the current circumstance.

Smiling is the most perplexing facial expression. Yes, some dogs grin, and the smile could seem frightful if you're not used to it. Dogs often show their teeth as a warning, as if to say, "Look at my weapons." A snarl is hostile, especially when a threatening growl accompanies it. The dog's front teeth are fully visible, and the corners of its lips form the shape of a C.

Dogs who smile also show their front teeth, but the intention is very different. This face, sometimes known as a subservient grin, often appears on a cheerful dog with a loose posture. The dog's demeanor indicates, "I come in peace."

Eyes

Your dog's eyes can reveal a lot about how they feel inside. The eyes of a dog can be either soft or hard. Soft eyes sometimes appear to be squinting due to their loosened lids. They suggest that the dog is content or pleased. Hard eyes, on the other hand, are when the eyes appear to become chilly. You'll recognize them when you see them as they signify a negative state of mind. The dog might be acting aggressively or protecting a toy. A dog that gives something a hard gaze, especially one that lasts long, typically indicates danger.

Dogs rely on eye contact as a crucial cue. Glancing away is meant to defuse a situation, just as the intense look may indicate impending aggressiveness. Dogs will purposefully glance aside and avoid eye contact when they're anxious. People often assume that their dog is being uncooperative or ignoring them when, in fact, the dog is just expressing discomfort.

Another important sign is the color of the eye whites. When a dog displays "whale eye," or showing the whites of their eyes, it's an indication that they feel concerned or anxious about something.

Understanding Dog Body Language

None of these dogs' nonverbal cues act alone. They come together as a whole. So, while interpreting a dog's signals, pay attention to everything it does, from the height of its tail to the shape of its eyes.

Your dog is always "talking" to you. You will establish a closer link of trust and respect if you understand what your dog says. Additionally, a better understanding of your dog's emotional condition will aid in behavior prediction and issue prevention.

SIGNS YOUR DOG NEEDS MORE MENTAL STIMULATION

Have you ever questioned why your dog continues to ruin your shoes, even after a two-mile doggy run? While many dog owners are concerned about their dog's physical activity level, they overlook mental activity's importance. If your dog is exhibiting any of the symptoms listed below, your dog may be in severe need of mental stimulation, such as training, activity toys, daycare, games, etc.

1. Can't Settle

Does your dog start to pace, whimper, or get up from where he lies every few minutes? Even after physical activity, dogs that simply can't relax deal with a brain that won't shut up. They'll ultimately calm down and relax after engaging in some mental exercise.

2. Destroys Things

No matter how many runs you give your dog each day, he probably must wear out his brain if he's still tearing, gnawing, shredding, and destroying your possessions. Give him a puzzle to solve or an action toy that he can "tear" into (hiding toys is fantastic!).

3. Tail chasing

Some dogs develop a tail-chasing obsession. However, any dog can develop into a tail chaser. Typically, it's herding breeds like the Border Collie. If you give your dog something else to think about (barring any medical issues), she will cease chasing after things. Things like trick training or teaching her to herd a ball into a net can help.

4. Barks at Everything

This dog acts as though she's "looking for trouble"; everything is a threat, even you opening the door or the fictitious rabbit in the corner. These dogs need to be trained, but it's also important to provide them with mental stimulation other than making up threats. Give them a task to complete, like telling them to locate their mat and lie down when the doorbell rings, to help them relax.

5. Digging

Has your dog dug a tunnel in the backyard that leads to China? Again, boredom is the usual cause of this conduct. Give your dog a brain-stimulating activity to engage in in the backyard, and they will be less inclined to destroy your petunias. Some pet owners provide a sandbox for their dog to dig in and bury objects for them to find—this requires mental and physical energy!

6. Sleeps A Lot

You might assume this is positive, but your dog may be bored if they do this, and it's undoubtedly unhealthy. You should get some activity toys if your dog sleeps more than your cat.

7. Whines

Your dog could be bored if he is whimpering for no apparent cause (you've ruled out pain, anxiety, tension, attention, etc.). How did you pass the time when you were bored as a child? You whined to your parents! Your dog is doing the same thing. Wait until he is calm, then give him something to do, such as an activity toy or play with him.

CHAPTER 3: INTRODUCTION TO MENTAL TRAINING FOR DOGS

How to Mentally Stimulate a Dog

If your dogs could talk, they'd regularly say, "I'm bored." But they are unable. Instead, they act inappropriately, destroy your stuff, and annoy you. To prevent this, dogs need to get brain stimulation. Keeping your dog's brain engaged is one of the most important things you can do to keep them happy, calm, and comfortable. Daddy and Mommy are also delighted when the dog is. Speaking about the young, several activities are also excellent for cognitively engaging puppies.

Therefore, you're in the perfect place to challenge your dog's intellect!

Have Your Dog Work For Their Food

Make your dog work for their food to keep their brains busy. Making them sit or lie down before you place their dish down or concealing little kibble across the house will help. Furthermore, you can also purchase puzzle feeders that test your dog's problem-solving skills.

1. Let Your Dog Explore and Sniff on Walks

One of the best things you can do to develop your dog's brain is just letting it smell. For dogs, sniffing is a terrific way to stimulate their wits and expend energy. Allow your dog to stop and sniff as much as he or she wants while you're out on a stroll. To reinforce the habit, you might even want to offer your dog a special reward when they stop to sniff.

2. Get Your Dog a Puzzle Toy

The best way to cognitively challenge your dog is with puzzle toys. They can be a terrific way to satisfy your dog's innate curiosity and keep their brain active. You can choose a puzzle toy from the wide selection on the market that is ideal for your dog's particular personality and demands. The Dogsee Play Treatoy is a well-liked puzzle toy. The best dog kibble or treats can be placed inside this toy, which motivates your dog to work to get the food out. To give your dog an even greater challenge, you can place peanut butter or any spreadable food inside the Dogsee Play Treatoy.

A wide variety of food-dispensing puzzle toys are also available. Your dog must use problem-solving abilities to learn how to get the food out of these toys.

3. Teach Your Dog Some New Tricks

While it's a common belief that an elderly dog can't learn new tricks, this isn't always true. You can teach your dog new skills that will help to keep their brain busy and engaged with a little perseverance and creative thinking.

Teach your dog to fetch objects from various areas of the home. This is an excellent way to cognitively engage your dog, as they must use their problem-solving abilities to locate the object and ascertain how to obtain it.

Use common objects to create an obstacle course on your lawn or home. Once more, your dog must apply its problem-solving abilities to learn how to navigate the course.

If you have more than one dog, teach them how to play tug-of-war. This activity can be an ideal way for dogs to socialize with one another and aid with cerebral stimulation.

Finally, try teaching your dog some basic skills, such as sit, stay, come, etc. This will keep their mind engaged and a wonderful approach for you to develop a close relationship with your dog.

4. Engage in some nose-work games

The best method to cognitively challenge your dog is through nose work. You can play this entertaining and simple game with your dog, and it's a wonderful way to strengthen your relationship. Furthermore, it's a great way to exhaust them!

You can play a variety of dog activities that include nose work. A common game is hiding puppy treats around your house and letting your dog locate them. This is a terrific way to put their innate scavenging tendencies to work, and it is a lot of fun for both of you.

Making a scent trail for your dog to follow is another nose work game. Rub an item with a strong aroma, such as an orange peel, with a piece of fabric to achieve this. After that, bury the material someplace within the home and let your dog discover it. This entertaining game is excellent for teaching your dog to follow its scent.

Nose work is a terrific way to challenge your dog and strengthen your relationship with them cognitively. So go outside and start practicing nose work with your dog right away!

5. Teach Your Dog the Names of Their Toys

One approach to stimulate your dog mentally is to teach them the names of their toys. This is a fantastic approach to stimulate their brain and catch their attention. You can begin by mentioning the item's name, followed by a command, such as "toy, sit." You can include more toys and commands as your dog grows more used to the game. This is an excellent way to exercise your dog's brain while training her.

6. Participate in Some Free Shaping Games

Shaping is a training technique for treating dogs when they exhibit the desired behavior. It's a lot of fun for both of you and a terrific method to teach your dog new skills or tricks. To get started, pick a habit that you want to shape. Anything from sitting down to putting their nose to your hand can be considered. Once you've chosen a behavior, encourage your dog by clicking or saying "yes" each time your dog does that trick. So make sure you say "yes" when your dog exhibits that behavior because you want to encourage the precise action you seek. Your dog will soon learn the actions you're after and will start displaying them more frequently with time and regular training. Playing brain games with your dog is an excellent way to challenge him and spend time together cognitively. Why not give shaping a try? It's a fun and simple way to get started.

7. Make an Obstacle Course For Your Dog

Using simple home items, you can make your dog an obstacle course. Your dog will like this brain training and have something to do. You only need some tasty, natural adult dog treats, patience, and ingenuity.

Here are some suggestions to get you started:

Use blankets or cardboard boxes to make a tunnel.

Make barriers for your dog to leap over using chairs or other pieces of furniture.

Use a blanket or towel and two chairs to create a bridge.

Give your dog some treats to find by hiding them around the room.

There are countless options! Just watch your dog as they play on the obstacle course. You don't want them to hurt themselves or damage your possessions.

8. Play interactively with your dog more often.

Playing interactively is essential for dogs of all ages, particularly puppies and young dogs. You can keep your dog's mind active and interested by engaging in interactive play. This is crucial for the growth and general well-being of their minds.

There are several ways to interact playfully with your dog. One approach is to teach them basic commands or tricks.

This keeps them cognitively active and deepens your relationship with your dog. Playing fletch, hiding goodies for them to uncover, or building up an agility course are other interactive games you can play with your dog.

Make sure you play interactively with your dog regularly, regardless of how you choose to. This will help make sure that their minds stay sharp and healthy!

9. Play the Shell Game With Your Dog

Playing the shell game with your dog is one of the finest ways to keep them intellectually active. Your dog will enjoy playing this game, and it will also help them develop their problem-solving abilities. You'll need three little cups, a small reward, and the shell game to play with your dog. Place the goodie under one of the cups, and mix the other cups. Let your dog explore the area and select a cup. If they select the right cup, they will receive the reward! You can increase the difficulty of the game for your dog by using many cups and hiding the reward in various locations each time you play.

If you're seeking engaging dog activities, the training techniques listed above are terrific for training your dog's brain while still having fun. Exercise is fun for your dog but can also benefit their mental health. Choose a brain training game or activity appropriate for your dog's personality and needs from the many available options. You can quickly observe your dog's behavior change with a little perseverance, work, and tasty training goodies.

CHAPTER 4: HOW MUCH EXERCISE FOR DIFFERENT DOG BREEDS

The greatest mistake in this situation is bringing home a dog who requires more mental and physical stimulation than the owner is prepared (or able) to give.

Dogs that don't exercise enough are in serious trouble. Exercise is well-known as "good" for humans but also crucial for dogs.

When dogs get enough exercise

If you can give an Appenzeller Mountain Dog enough rigorous activity, they make excellent family pets. If you can't, he is not an amazing family dog!

Expect easier training. Dogs who have had enough exercise are easier to train because they are happier, calmer, and more able to focus on you.

Expect to see fewer behavioral issues. Well-exercised dogs get into less mischief because weary dogs sleep.

Expect to feel better. Exercise helps your dog live a longer, more comfortable life by keeping his heart, joints, lungs, and muscles strong and healthy.

When dogs don't exercise enough

Be ready for behavioral issues. When dogs don't receive enough exercise, they often let out their pent-up energy through destructive chewing, barking, or digging. Many dogs who are labeled as hyperactive exhibit these behaviors because their owners don't exercise them enough.

Dogs are lively, intelligent animals. They become bored and irritated if all they can do is walk about the home or yard aimlessly. Dogs can trash objects, bark, or act rambunctiously to express their restlessness and dissatisfaction.

Bored. Bored. Bored. In addition, they are free in the house while no one is at home, a luxury they have not yet earned!

When a dog's needs aren't being met, getting him to behave the way you want him to is more difficult. Yes, you must stop him from damaging things and barking, but you must also give him more exercise and engaging pursuits.

Get ready for health issues and a shortened lifespan. Your dog's heart, lungs, and joints weaken from lack of activity, increasing their susceptibility to disease. Lack of activity also contributes to weight growth, and overweight dogs are more susceptible to digestive, musculoskeletal, and cardiovascular illnesses and diabetes.

The lesson is to avoid getting a dog requiring more activity and mental stimulation than you are prepared (or able) to supply.

The misconception about "exercising" in the backyard

Some owners believe they have a perfect answer to the exerciseproblem.

Are you expecting your dog to receive all the exercise he requires by "running around" your yard?

That may be the case if you adopt a small dog, an elderly dog, a dog with a disability, or a dog with health problems that prevent them from handling a lot of exercise.

For most other dogs, "exercising" in the backyard doesn't happen. Most dogs simply take their time and lazily amble around, sniffing, picking up a stick, sniffing, dropping the stick, sniffing, peeing, sniffing, pooping..., and then sprawling out in the sun (or shade) to nap.

Fence running is detrimental psychologically even if it is considered "exercise."

Some dogs "exercise" in the backyard by sprinting along the fenceline while barking at onlookers, UPS trucks, the neighbor's dog, or the neighbors themselves.

These dogs aren't having a good time. They are spewing out of boredom and frustration.

Can TWO dogs keep each other exercised?

Could be. If you have several dogs that chase, play, and wrestle each other, a lot of energy will be expended. Simply include some vigorous walking, ball play, or stick retrieval.

However, just because you have two dogs doesn't mean they will automatically exercise one another. Many dogs that live together actually do so together. Even if they might follow one another around, that is not exercise.

Now, if YOU go outdoors and walk (or run) about the yard with your dog(s), he/they will receive much more exercise, especially if you play tug or fetch games.

What is mental stimulation or exercise?

I'll begin a sentence, and I think you can finish it for me. Ready?

Most dogs were developed to do some sort of...

(yes!) WORK. Good work!

Breeds with a history of labor-intensive work demand a lot of physical and mental exercise.

Chesapeake Bay Retrievers

Chesapeake Bay Retrievers were created to spend the entire day bringing shot geese back from the bay's frigid waters. Toughness, stamina, and endurance are three qualities that make this dog an excellent working dog, but it is not an excellent pet unless you are ready to give it a ton of mental and physical training.

Providing the dog with mentally stimulating activities is what I mean when I say "mental exercise." Many breeds get bored, irritable, and inclined to express discontent via mischief and behavioral issues without mental and physical activity.

A mini-obstacle course in your yard or mentally challenging activities like hide-and-seek, in which you hide, and your dog searches for you, or find-the-hidden-toy are some options for entertainment.

Something to do might include backpacking/hiking trips, hunting, swimming, and fetching sticks at the lake, or guarding livestock on your farm.

Attending regular courses at a nearby dog club where you may study advanced obedience, herding, scent-work, agility, tracking, weight-pulling, carting, protection sports, or sledding could be something to do.

Samoyed

The Siberian Samoyedic peoples were nomadic and relied on Samoyeds to tow sledges and herd reindeer. Samoyeds were present when the polar expeditions to the Arctic and Antarctic were equipped. It should come as no surprise that sporty outdoor enthusiasts would want to possess this resilient, adventurous breed.

More than HALF of the 200+ breeds that are easily accessible in the US have histories as working animals and require a lot of mental and physical activity (unless you adopt an elderly or disabled animal). Can you see why mixed-breed and crossbred dogs are such excellent options when looking for a pet?

Breeds that demand the LEAST amount of exercise

MINIMAL does not include strolling outdoors to potty and returning inside to sleep!

If your dog is a game, play fetch every day.

Make a little obstacle course and run around it with him while praising, rewarding, and persuading.

Teach tricks.

Take walks while the weather is beautiful. (The majority of these breeds detest cold or rainy conditions.)

Maltipoo

While not requiring a lot of exercise, this mixed-breed Maltipoo won't reject it if you offer more! Just like bigger dogs, even tiny dogs enjoy playing and running about.

Tiny dogs who do okay with MINIMAL exercise:

Papillon

Affenpinscher

Japanese Chin

Brussels Griffon

Yorkshire Terrier

Pomeranian

Miniature Pinscher

Chihuahua

Toy Fox Terrier

Toy Manchester Terrier

Toy Poodle

Maltese

Dachshund

Boston Terrier

Boston Terriers don't require a lot of exercise, but most of them are energetic and fun-loving when they are young or middle-aged. Seniors are more subdued. Exercise this breed sparingly in hot weather as their deformed facial structure causes respiratory problems.

Small dogs who respond well to VERY LITTLE exercise:

Lowchen

Bichon Frise

Pekingese

Lhasa Apso

Coton de Tulear

Boston Terrier

Dachshund

Tibetan Spaniel

Shih Tzu

Pug

Havanese

Medium-size and midsize dogs who do okay with MINIMAL exercise:

English Bulldog

French Bulldog

Breeds that need LOTS of exercise

Here are several dog breeds that require more exercise than minimal. These breeds require a lot of physical and mental activity because they were developed to work all day.

Start with two daily 45-minute BRISK walks, where BRISK means the dog should be trotting rather than merely walking slowly.

Provide vigorous activities such as running in a secure enclosed space, fetch games, tug of war, obstacle courses, hiking, carting, swimming, jogging, or any other activity the specific dog likes every day, or at least every other day.

Small and medium-sized dogs that require A LOT of exercises:

Jack Russell Terrier

Border Terrier

Lagotto Romagnolo

Fox Terrier

Border Terrier

Border terriers have a lot of drive. While they are not overly active but determined to engage in demanding activities. Owners must offer engaging activities; otherwise, the Border will find them alone!

Medium-size dogs who need a LOT of exercises:

Australian Kelpie

Australian Cattle Dog

Border Collie Portuguese Water Dog

Vizsla

Australian Shepherd

English Springer Spaniel

Brittany

Nova Scotia Duck Toller

Siberian Husky

English Shepherd

Australian Shepherds are action-oriented dogs. Interesting stuff. Lots of interesting stuff! If that doesn't seem like your family, either get an older, more subdued Aussie or go with a different breed Young and middle-aged Aussies want to do many exciting things, after all!

Giant and large dogs who need A LOT of exercises:

Alaskan Malamute

Airedale

Dalmatian

Appenzeller Mtn Dog

Flat-Coated Retriever

Beauceron

Giant Schnauzer

English Pointer

American Bulldog

Coonhound

German Shorthair

Belgian Shepherd

Briard Irish Setter

German Wirehair

Gordon Setter

Cane Corso

Bouvier des Flandres

Catahoula Leopard

Weimaraner

Labrador Retriever (field)

Spinone Italiano

Curly-Coated Retriever

Bringing home a dog whose exercise needs are far greater than the owner's desire or capacity to care for is one of dog owners' biggest blunders. The breeds mentioned above are not for couch potatoes!

Breeds that require SLIGHT exercise

Which breeds require a little exercise? Any breeds for which I have written reviews but have not included under MINIMAL or LOTS.

All dog breeds, on average, require moderate activity.

Start with two daily 20–30 minute BRISK walks, where BRISK means the dog should be trotting rather than merely walking slowly.

Add vigorous exercises like running in a secure enclosed space, retrieve and tug games, obstacle courses, jogging, carting, hiking, swimming, or any other activity that the specific dog loves as often as you can.

Cockapoos are active, bright crossbreed dogs who love playing,romping, and doing fun activities.

Sighthounds require a particular kind of exercise:

Interested in one of these breeds? Pharaoh Hound, Borzoi, Ibizan Hound, Afghan Hound, Greyhound, Irish Wolfhound, Italian Greyhound, Saluki, Whippet, Scottish Deerhound?

They are all members of the sighthound breed, a group of tall, thin, long-legged hunting dogs. Their task is to detect even the smallest movement of prey across great distances and follow it with incredible speed.

Sighthounds don't require long-distance running. Sprinting is more their thing than endurance. The only place they can suddenly burst into an enormously strong, driving, floating gallop that no other dog breed can equal is a sizable, securely walled yard.

However, after a few vigorous laps around the yard, a sighthound is ready to give up and nap on your plush couch—at least until later, when he is eager and ready to do more laps.

Therefore, sighthounds shouldn't be kept in a small yard or one that isn't securely fenced. Electronic/invisible fencing? No, they will plow through any "shocks" and catch and kill their hapless prey before leaving. This happens whenever they notice a running creature.

CHAPTER 5: SENSORY STIMULATION

Your family may adore your dog, but it's typical for people to underestimate how smart their dogs are. This means that it's common for households to forget to engage their pets in enrichment activities.

If your dog looks bored, anxious, or restless, it probably needs additional enrichment activities. Even while you presumably already walk them daily, they might benefit from various activities, particularly ones that engage their senses.

Why Dog Enrichment?

Dogs aren't meant to lie around all day waiting for their owners to get home. They are built for action instead. When you enrich your dog's environment, it boosts both their psychological and physical well-being.

One of the various enrichment methods is the use of sensory activities. These are any pursuits intended to stimulate one or more of your dog's five senses.

Here are some suggestions to stimulate all of your dog's senses.

Smell

Dogs like utilizing their sense of smell to discover and comprehend their surroundings. You can excite them by regularly exposing them to different odors.

For example, do you or someone you know keep goats or other animals? If so, you might consider leaving an old T-shirt in their living space for a few days. Bring the t-shirt home afterward, and fasten it to a tug toy. Your dogs will never be able to get enough of that t-shirt.

Hide some kibble in a blanket or snuffle mat for your dog to sniff out—this will also enhance their sense of smell. Do you not know what snuffle mats are?

Snuffle mats are ideal for challenging your dog's intellect while rewarding them with delectable goodies. They are constructed from a rubber sink mat that has fleece strands tied to it. Snuffle mats are available both ready-made and homemade.

If you hide a few goodies between the fleece strips, your dog will be absorbed in trying to find them right away.

Remember that your dog's ability to smell is very important. It's how they prefer to learn about the world; therefore, stimulating them frequently is important.

Touch

Physically touching people, an item or another animal may be soothing for dogs. It's crucial to give tactile sensations, such as those associated with caressing and hugging, but even something as simple as brushing your dog's coat can be beneficial.

Another choice is to walk your dog along a grated or rough surface. To ensure your dog's comfort and safety, watch out for any nails or other sharp items and make sure the grating isn't too widely spread.

Your dog could require time to engage in digging, mouthing, and pawing as he moves over different surfaces.

Bodies of water can also be a great environment for your dog because they offer a variety of tactile experiences.

Your dog will feel more at ease after returning home, thanks to these tactile sensations.

Taste

Dogs' oral senses provide them with a wealth of stimulus and information. Simply providing a range of snacks regularly is one of the best ways to do this. Who, after all, enjoys the same sweets every single time? Ensure you have a variety on hand to provide your dog variety and keep things exciting.

Another source of oral stimulation is a plush Kong toy. Many dogs enjoy this, especially those who enjoy working for a nice reward. Puppies can benefit from this pastime in several ways, including stress reduction and life enrichment.

Utilizing food-dispensing toys is an additional choice. You can make yours with some PVC pipe, a concept from zoos. Even the dog with the most persistent chewing behavior cannot destroy these toys if they can withstand tigers and bears.

Just put two end caps on a piece of food-grade PVC pipe. One of these caps must have a screw top so you can easily take it off and put it back on. Make a few holes in the PVC pipe with a drill that is just slightly bigger than your dog's kibble. This puzzle-like pastime is a favorite among dogs because it satisfies their chewing urge.

Don't forget to provide your dog with various chew toys. Toys with knobs, spikes, and other tactile features will be more fun to chew on, so watch for such.

Sound

There are many good reasons to use various sounds to excite your dog. Dogs might be frightened by strange sounds, one of the main causes.

Simply dumping a bowl or piece of flatware on the floor will help your dog become accustomed to loud, unexpected sounds so that he won't be as frightened when actual accidents occur. Several dog owners swear by utilizing a blow dryer for sound stimulation.

Another option is to get some wind chimes. Several dogs adore the tranquilizing sounds.

Early exposure to various sounds can significantly impact how your dog will respond to various sounds in the future.

Sight

Do you take your dog for walks on the same path every time? Being used to something can be enjoyable as it helps you and your dog observe changes over time. However, exposing your dog to various situations will undoubtedly be beneficial.

Find fresh locations to walk your dog regularly. There are many fresh views and fragrances to experience on a new path or street. Work on visual signals with your dog when you are at the dog park or out on a walk to further pique their visual sense.

CHAPTER 6: SOCIAL INTERACTION

How to Socialize a Dog with Other Dogs

It's crucial to socialize your dog with other dogs. Particularly if you're thinking about introducing a second dog into your house, you want a dog to be able to get along well with a variety of dogs. Socializing with an older dog can be more challenging than with a puppy. You must begin with small interactions with other dogs. Be sure to reward good conduct and deal with bad behavior correctly. While an adult dog may never be entirely sociable or playful with other dogs, you can teach an adult dog appropriate behavior around other animals with little effort.

1. Exposing Your Dog to Different Dogs

Begin with an interaction with a single dog. A socially mature dog, or between one and three years old, won't like playing with many other dogs if it has previously been socialized. Your dog would feel overwhelmed in a place like a dog park. Start with brief, regulated interactions with a single dog. Find a family member or acquaintance with a well-socialized dog, and take your dogs on a walk together. Locate a buddy who has a calm, relaxed dog. Have the dogs meet in a public place, then stroll together while maintaining a safe space between them.

If the dogs behave during the walk, allow them to engage with one another and gently smell each other. If aggressive conduct happens, talk to both dogs calmly until they stop acting aggressively.

The two dogs may eventually reach a stage where their tails and bodies begin to wag whenever they are around one another. At this point, you can allow them to interact while off-leash in an enclosed yard.

Walk on various paths to meet new dogs. Introducing your dog to various animals and circumstances is a crucial part of socializing. Your dog's experiences will be confined if you walk him on the same path daily. Change your walking path daily to expose your dog to new sights, sounds, and dogs. Experiment in a wide range of situations. A dog must maintain composure and good manners in a variety of settings. Try sidewalks, dirt roads, walking paths, and concrete. Vary the amount of activity. One day, walk your dog through a peaceful neighborhood; the next, take him through a bustling area of your city.

Drive to a whole other neighborhood on some days if you can.

If it's a safe choice, consider a dog park. Visiting dog parks can greatly enhance a dog's chance to socialize. Your mature dog, however, could not have received any prior socialization. In this instance, a dog park might be concerned about it. If you know your dog was socialized as a puppy, taking it to the dog park could be beneficial. If you got your dog from a shelter, you might not know much about its past. If you're unsure whether your dog has ever been socialized, observe its body language when you're in a dog park. Until you gauge your dog's reactions, keep it on a leash at the park. If your dog keeps close to you, growls at any time, and avoids other dogs, a dog park is probably not safe for your dog. Your dog could not have had much experience socializing with other dogs and detest being around lots of them.

You should avoid using dog parks as a socializing venue in this situation. Keep your interactions with specific dogs under control.

Consider this: if you were really shy, you would not want to be in a room with 35 other people, would you? The same principles apply to timid dogs.

If your dog starts acting scared or aggressively, remove them from the situation. It's common to think of aggressive actions like snarling, growling, and barking as a strategy to assert authority. As a result, many individuals tend to penalize certain actions. These actions, nevertheless, are motivated by dread. If your dog acts aggressively while socialized, remove it from the environment and calm it down. This is more effective than punishment. This is more effective than punishment.

Try to divert your dog's attention when your dog barks at another animal. Use food or toys as a distraction, or call your dog by name.

Take your dog out of the situation. Get it far enough away from the other animal to be safe. From now on, speak to your dog reassuringly until it begins to relax. Return your dog to the issue once she has calmed down.

HOW TO SOCIALIZE A DOG WITH HUMANS

An unsocialized dog poses the risk of acting viciously toward otherdogs, animals, or humans.

Fear causes dogs to become violent.

Through socialization, there is no longer a chance that your dog can develop a phobia of something that isn't genuinely scary, like unfamiliar people.

Due to this, understanding how to socialize a dog is crucial.

It is best to socialize dogs when they are four months old or less because it becomes much more difficult for them to develop habits after this point.

Dogs can develop phobias towards strange people, dogs, and sounds if they are socialized as puppies but never again as adults.

It is still achievable to socialize dogs that weren't socialized as pups, however it will take longer.

Although you can expect your unsocialized adult dog to get less hostile as a result of adult socialization, don't expect a dog that is suddenly welcoming as adult dogs' personalities are rather hard-wired.

My point is that teaching your dog socialization skills while he's young is one of the best things you can do for yourself and your dog.

How Do You Start Socializing Dogs?

Avoid separating the dog from the litter before the puppy is 8 weeks old as this is the first stage in socializing the dog.

Dogs wrestle, play, and interact with their mother, siblings, and other dogs during those eight weeks.

Those eight weeks are when dogs learn the most about getting along with other dogs, so removing the dog too soon might have long-term effects.

Additionally, letting your dog reside indoors can help it develop social skills.

Dogs won't be able to socialize as much as they should when they are always left outside.

Cummings School of Veterinary Medicine claims unsocialized dogs have a 580 times higher risk of developing violent behavior.

An aggressive dog has experienced fear. The best way to help a dog overcome whatever fears it may have is through proper socializing.

A dog won't react fearfully whenever a stranger enters the room if you train them to be around other dogs and humans.

Regular walks greatly enhance a dog's sociability. When you take your dog on a stroll, she will get acclimated to seeing unfamiliar people and dogs.

They could first feel nervous, but they will rapidly get over that anxiety if they walk every day.

Dog parks are a great place to socialize your dog as well.

If your adult dog finds it difficult to socialize well, you should attempt alternative socialization techniques, such as leashed walks, before bringing your dog to a dog park. You don't want to take the chance of your dog biting another dog.

However, if you have a puppy or an older dog that has already been socialized and merely requires a refresher, the dog park is excellent.

The opportunity for interaction and play with other dogs is extremely helpful for socializing your dog.

Introduce your dog to the other dogs and their owners at the dog park.

Your dog won't grow terrified when they meet more strangers when they have met people of diverse sizes, shapes, and looks.

Another great location to expose your dog to other dogs is at the pet supply shop.

This one may seem a bit strange at first, but it works. While many dog owners don't bring their pets to the pet shop, others do.

Your dog will meet new people and discover new smells. Additionally, it doesn't take much effort to take your dog to the pet shop.

Dog Aggression Warning Signs

When startled, even the kindest dogs might become aggressive.

Dogs are extremely fearful of noise, which makes them hostile.

Introduce as many noises to your dog.

Consider any loud noises that can startle your dog, such as those made by cars and vacuum cleaners, and expose them to such sounds as often as you can.

Over time, Fido will develop acclimated to such noises and won't really care anymore.

Some situations could terrify a dog and propel them to lash out.

Exposing your dog to these situations is the best way to help him become acclimated to them.

Your dog may get fearful if the collar holds them, touched on the underarms or paws, or have their nails trimmed.

In addition, if someone approaches your dog's food bowl, they can get hostile. The more of these experiences you expose your dogs to, they will be able to maintain their composure when necessary.

While it may be used on older dogs as well, this training technique is most effective when used on pups.

With this type of training, be more careful and gentle with mature dogs. Giving your dog treats while they remain quiet helps hasten the formation of a habit.

Your dog can become aggressive against others if you are the only human there.

Dogs can also feel afraid of people who don't look like the people they are familiar to.

You should make an effort to socialize your dog with as many different types of people as you can.

Ensure to introduce tall people to your dog if they are used to shorter people.

If they are used to thinner people, Introduce them to larger persons, and so on.

Even if you don't have kids, it's important to socialize your dog with them.

Children could approach your dog without asking first, and if he gets scared, he might bite one of them.

If your dog is comfortable with kids, you won't need to worry about that risk.

When learning how to socialize a dog, remember that dogs also experience fatigue.

Playing with another dog can satisfy your dog fully, but only until the other dog starts acting inappropriately. The opposite might also take place.

Dogs might get violent if they feel irritated by your dog after they have had enough playtime.

When you're out socializing your dogs, limit the amount of time your dog plays with other dogs.

Ensure you are familiar with your dog's personality and watch for any indications of fatigue.

Once you see the warning signals, it's time to return your dog to his or her home.

What To Be Cautious Of

As crucial as socialization is for your dog, safety must always come first.

Knowing the warning indications of a dog's anxiety or discomfort is the first step in exercising caution.

Be sure you comprehend the habits exclusive to your dog because each dog's personality is unique.

Most dogs will yawn, pant, and place their tail between their legs to indicate they are about to become aggressive.

Dogs who are feeling agitated may shake their fur.

If the hair on their back stands up, they may be intimidated.

Remove your dog from the situation as soon as you see these signs in your dog.

This will prevent your dog from hurting himself or other people.

Be wary of the dogs you introduce your dog to when you take them for dog socialization.

If you have a small dog, she shouldn't be allowed to approach larger, unfamiliar dogs.

Because you can never be sure of the other dogs' reactions, your dog is put in risk.

Introduce your dog to dogs of a similar size. Get to know the other dog before introducing yours to one that is larger or smaller than it.

Ask the owner about the dog's personality to learn more.

Introduce the two if you think they might get along, but keep them on their toes during the conversation.

Once you're convinced the dogs get along well with one another, only let your dog go off leash.

If you think that socializing your dog is ineffective, consider training courses. This is more likely to be required when it comes to older dogs with ingrained, enduring behaviors.

Training sessions will provide your dog new opportunity to interact with other dogs in a controlled environment and will teach you about your dog's behavior.

If you are struggling to socialize your dog on your own, there could be something about your dog that you don't understand.

With the aid of an experienced trainer, you can pinpoint the root of your issues and resolve them.

Learning how to socialize a dog could seem tiresome; after all, is it really that crucial?

The brief response is that it is extremely necessary.

The appropriate form of socialization keeps the people around your dog safe and stops the emergence of future problems.

Additionally, a well-socialized dog won't start a fight with a strange dog that is (potentially) stronger.

Unfortunately, aggressive dogs are often euthanized. Therefore, keeping your dog safe requires dealing with any aggressiveness issues.

A dog should be socialized as much as possible while they are young. However, with a little more time and effort, you can socialize a dog after they are an adult.

Regular socialization exercises like walks are a great way to ensure your dog maintains puppy-like social skills.

The key is exposure. The more exposure your dog has to other dogs, people, social situations, noises, and other things, the better your dog will respond.

The easiest way to reverse socialization is isolation, so try to keep your dog as far away from that as you can.

Always be aware of when your dog starts acting fearfully, defensively, or aggressively. While each dog exhibits various indications, they all generally exhibit similar body language.

Never disregard symptoms like excessive panting, tail tucked between the legs, hair sticking up, yawning.

When you see these, take your dog out of there right away.

When learning how to socialize a dog, your primary concern should be ensuring the safety of your dog, other people, and other animals.

CHAPTER 7: BEYOND THE BASIC COMMANDS

You can speak to your dog by issuing commands. She learns the rules and what is expected of her via commands.

Both physical signs and verbal commands are valid modes of command. Dogs are able to pick up on our tone and nonverbal clues.

When taught the meanings of words, dogs can pick up a large vocabulary.

To prevent confusing the dog, it's essential to continuously employ the same words—cues.

You can also develop consistent hand gestures. Hand gestures may be useful because they typically motivate a dog to focus more intently during training.

They are also essential for training deaf dogs.

One of my older Shelties lost her sense of hearing when she was about 14 years old.

Without a leash, I could take her wherever and be certain that she would answer because she was aware of hand signals and kept an eye out for them.

Training dogs is similar to building bricks. A dog's knowledge expands with each passing day.

Why Should My Dog Be Trained?

Training will bring you and your furry best buddy closer together. It aids with rules and standard learning as well. That gets him ready to do as you say.

Living with a trained dog is enjoyable! He can go with you to the pet store, the beach, or a friend's home.

You won't be pulled down the street by anybody. Your friends won't trip over when they come into your house, so you'll see them often.

There won't be a closed door between you and your dog while you dine. No one will use your couch as a chew toy.

Furthermore, training supports the development of impulse control in your dog. He will get more used to the demands and eventually develop some self-control.

When your dog has impulse control, they may perform well in a range of circumstances.

Commands used in dog training are especially beneficial for young pups, who may lack impulse control.

Dog training provides a lot of advantages!

Basic commands will be quite helpful in your day-to-day activities. Every dog should know a few commands.

Once your dog has mastered the fundamental commands, the possibilities are endless!

The list that follows starts with basic instructions like sitting and paying attention before moving on to more difficult instructions like take it/hold it.

Your dog can be trained to do tricks like speaking on command.

Dog Training Commands

Before we start, I want to make two suggestions.

- Get some expensive treats for the dog's training.
- Get a treat pouch.

1. Attention Command

All of your encounters with your dog start with getting his attention. You are just commanding the air if he is not paying attention to you.

The Attention Command: How Do I Teach It?

When you call he name and she turns to face you, say "yes!" and give your him a small treat.

Hold the treat near to your face and show him first if he won't look at you. Call him name and give her a treat if he looks.

Repeat the practice numerous times. Do not bore him.. End while he is still joyous and jubilant.

The reward doesn't have to be held up to your face the whole time. Once he realizes that gazing at you brings rewards, say the name, give the order, praise him, and give him a reward.

You can add "look" and "watch" instructions once he discovers that gazing at you is rewarding.

2. "Come" Command

The come command is one of the most crucial orders you can give your dog. Your dog's life can be saved by a reliable recall!

Everybody is familiar with dogs that flee when called. It's awkward and annoying.

The chasing game will go their way. We only have two legs, but they have four.

The Come Command: How Do I Teach It?

Show your dog a high-value treat at first. Even resistant dogs can be persuaded to come when offered a treat they can't refuse. Hot dogs, steak, cheese, and stinky fish treats worked well with my dogs.

Use something that he truly enjoys. Then call his name and say happy, "Come!"

When teaching this to your dog, you can show the treat as bait before uttering the command.

First, start by leashing your dog. Use a long line once he learns it. The longer leash prevents him from simply running off. Start without distractions, as with other commands.

Mark the conduct as soon as he approaches you. Say, "Yes, come!" Give him a jackpot of delights. You need to make coming to you more fun than all the other things that may divert them!

Once he starts showing up on a regular basis, solely provide the reward treat and cease offering the lure goodie.

To ensure the dog wants to come when I call him the next time, I normally reward him with three or four tiny treats once he comes. It is known as a jackpot!

When he discovers how much fun this is, he'll start to show up regularly. And you'll feel less stressed.

Don't call your dog over to scold him for inappropriate behavior, such leaping up on the counter. Teach him to "leave it." In this case, even placing him in a yard he enjoys or coming in from the yard can be detrimental. His fun is over.

Dogs are perceptive creatures, so if anything negative occurs after they come to you, they will learn not to come the next time.

He may have learnt not to respond to the word "come," but that's no reason to give up. You may get some practice by replacing the word with a different word, such "here." Be consistent in your usage of words.

Play games with him within the home while calling his name and praising him when he comes running.

A dog can have much fun playing the "find me" game. You don't want to induce separation anxiety, so don't play it too frequently.

3. “Leave It” Command

Everybody has encountered a dog who seeks out trash and picks up food wrappers while walking. Or the enticing roast sitting on the counter.

It is annoying and can result in a sick puppy getting into the wrongthings.

Therefore, you need to train your dog to obey your "leave it" command by never attempting to pick up the forbidden object.

How Do I Teach it?

Clutch a snack in your closed hand and say, "Leave it." Don't rush him.

Say "yes!" and reward the dog with a treat when he removes his nose from your closed fist.

He will learn to keep attempting to steal treats from your hand if you give him the one he was sniffing. So don't give him that treat.

Another way to teach your dog this command is to use a six-foot leash and hold it loosely enough for your dog to feel comfortable.

A treat should be thrown about six feet away, well beyond his range. Ensure that he sees you throw it.

Say, "Leave it!" as it lands. Be patient. When he relaxes his grip on the leash and stops tugging in that direction, it's time to say, "Yes!" excellent, leave it" and give him a treat.

To avoid him learning how to acquire the treat you wish him to leave alone, you should not pick it up off the floor.

Once he understands that he shouldn't pick up the treat off the floor, you may go on to other items, like the TV remote, your shoes, or your brand new smartphone, or other things that you don't want him to get down with.

You want him to always comply with your requests to leave.

You'd want it to sink in that there are restrictions on a range of heights for some kinds of things, too. You can teach him by setting up the things on a table or counter and doing the exercise as described.

BONUS TIP: We begin by using their kibble as treats while teaching our pups and dogs. However, you'll want to offer a greater reward in some situations.

4. “Sit” Command

Most people start their pets out by teaching them this command.

How To Teach the Sit Command?

When your dog has had enough activity, gently slide a reward backward while holding it directly over his nose. Say "Yes!" softly in response as his behind touches the ground. "Good sit!"

If he leaps for the treat, ye may be overexcited or you are putting the treat up too high.

5. "Down" Command

You can train your dog to lie down after he has mastered the sit command.

The "off" command for jumping is different from this one. It is lying on his stomach and legs, flat on the floor.

If your dog already lays down when you tell it to, you can teach it to stroll gently to its resting spot, such as a dog bed.

If you take him with you to a restaurant outside, he won't steal food. He won't be able to climb on top of anybody or anything either.

How Do I Teach the Command?

Ensure your dog has initially engaged in enough physical activity. Sit him down, then place a reward just in front of his nose and slowly lower it toward the ground.

Hold off till he relaxes and lays down.

If he rises up or crouches, you may keep the reward sliding down too quickly, or he becomes too active.

Say "Yes, good down" calmly as soon as his body reaches the ground.

ADVICE: He'll pop back up if you compliment him too enthusiastically.

6. Loose Leash Walking

It's crucial for both you and your dog to learn how to walk on a loose leash.

You run the risk of being hurt and losing your dog if you accidentally let go of the leash.

One of my clients had been pulled over by her dog, breaking her wrist. Then, she hired me.

She phoned me out of panic and didn't want to find her dog a new home. She shed a few tears as she explained what had occurred to me.

We trained her dog to walk politely while being leashed. After that, they could enjoy taking strolls through the neighborhood. The dog also received extra exercise.

Loose Leash Walking Command - How Do I Teach it?

Exercise your dog before practicing, if possible. Play fetch. To get his head going, perform some other training activities first.

Given that you're likely to walk your dog so that he exercises, it can seem silly to exercise your dog first.

However, taking this extra step comes before your training sessions. He is being prepared for success.

Between your training sessions, you can continue going for your usual walks. The leash should be short and have some slack.

Choose a side. Your dog should always walk on your left or right side, depending on your preference.

Have him stand or sit next to you. Reward him by giving him a treat when calmly seated beside you.

Decide what you will command. "walk" and "Let's go" are often used.

Name your dog, then give the command. Move closer to him while holding a goodie in your hand. When he remains close to you, give him the treat.

Take more steps, and reward him when he stays by your side.

If he begins to pull, give up and wait your time until he returns to you. Then restart.

To help him learn not to pull, you can make about turns and move oppositely. When he walks close to you, compliment and praise him.

Anti-pull tools like no-pull harnesses can be useful. I've successfully adjusted Freedom No Pull Harnesses, Easy Walk No Pull Harnesses and other comparable products. These two harnesses work well with dogs like labs and Dobermans, whose legs are proportional to their bodies.

These harnesses may not fit dogs whose bodies are long and have small legs. An anti-pull harness like the Easy Walk can be quite helpful for dogs with longer backs and shorter legs like Shih Tzus. Larger breeds like labs can benefit from the Sporn as well.

Mark the desired behavior with a reward and say, “Yes! Good, let’s go!”

7. “Wait” Command

If you give him this command, your dog may not dart out the door. Most dogs want to run outdoors when a door opens because they are thrilled to see the excitement.

However, this may be devastating. Your cherished pet can wander off or run into traffic.

These nightmare scenarios won't happen if he is taught to wait and not run outside or even to wait at curb while he is walking.

And you'll enjoy your walks more.

Different trainers employ the word "wait" in different ways.

It's typically employed as a break during which the dog doesn't advance but, unlike a stay, is not required to stay in one place.

Thus, you can use it to keep your dog from getting out of your vehicle by telling him to "wait" before putting on his leash and then telling him to get out of the vehicle.

How Do I Teach the Command?

Your dog needs to have some slack on a short leash. Give a "Wait" command. Once your dog stops tugging and comes to you, give it a treat and some praise.

Before continuing, ensure he has calmed down for at least a fewseconds. Make him wait for as long as he can before continuing.

Give him your motion cue before you go on, such as "let's go."

After a few successful attempts, repeat this exercise a few more times before moving on to a different one. Alternatively, end the training session and reward your dog by playing with him!

8. “Settle” Command

This command is another impulse-control exercise that might improve your relationship with your dog. Like us, dogs can become overexcited.

However, unlike people, they typically leap, rip clothing, and cause mayhem when overstimulated. Getting adequate exercise before practice is beneficial. When they know what it means, they can "settle" on command.

It has been my observation that training a puppy to settle down may greatly enhance the dog's ability to regulate its impulses.

My Lhasa apso dog was often overstimulated by a variety of activities when he was a puppy.

He found it quite easy to calm down after learning the "settle" command, so he started acting more composed.

Make sure your dog has enough room to spread out comfortably on the bed you use to let him relax.

How Do I Teach the Command?

Put your dog on a loose leash and tell him to "settle" while you gently reward and praise him and give him a little goodie.

This is not the same as instructing him to lie down on command. Every time the dog remains quiet throughout the "settle" training, he should receive praise. You want him to realize that maintaining composure is what is being appreciated.

If the treat lands close enough to him and don't cause him to stand up, you can throw it his way.

Eventually, he will settle on command if you say "settle."

9. “Stay” Command

When your dog has mastered the "sit" or "down" commands, you can begin working on the "stay" command.

Stay means remaining still in one's current location without turning or moving up.

He has to learn not to creep up on people or edge away.

How Can I Teach the Command?

Teaching this involves the "three D's"; distraction, distance, andduration.

It's better, to begin with duration first. Put your dog in a sit or down position close to you and command him to "stay."

As you give the order, you can also place your hand in front of him with the palm side down and the fingers pointing downward to further emphasize the "stay." Expect him to move after a short while.

Build time gradually throughout workouts. It pays to be patient.

To ensure that the dog knows what you want rather than merely expecting the requisite amount of time, alternate between shorter,longer, and shorter durations.

Make sure you release the dog from the stay. If there is no beginning or end, staying has no purpose.

You can teach him a word to say when standing up is okay. Utter a word like "break."

Say, "Good break," as he gets up, but don't make it seem like more enjoyable than the actual stay.

You may begin to move away from him until he comprehends and learns to stay, then you can return and set him free.

As he can handle it, increase the distance. Never call the puppy; always come back and let him go. You can introduce distractions when he proves his reliability.

10. “Take It” And #11. “Hold It” Commands

Use this command to train a dog to fetch. It can be used for tasks like getting a Coke can or a television remote.

The dog may be taught to execute other tricks that include him or her picking up and holding items. My golden retrievers were among the many dogs I taught this to.

I wanted them to take things when I told them to, even if they naturally retrieved them.

Take It and Hold It Commands - How Do I Teach It?

Hold a beloved toy within easy reach. He should be compelled to grab it right away. Try to make it more enticing if he doesn't by swinging or squeaking it back and forth.

As soon as he holds the toy, tell him, "Good, take it." However, he will drop the toy if you offer him a small treat.

Encourage him to hold onto the toy for extended periods. Once he has mastered different toys, have him grab more safe things.

12. "Give" or "Drop It" Commands

Everyone is familiar with a dog who tugs relentlessly. He snatches your hat and flees with it. His jaws clamp down as you try to take it, tearing your hat to pieces.

That issue can be resolved by teaching a dog to give something up, which can also educate them not to guard the item.

The Drop It or Give Commands: How Do I Teach It?

Once you have your dog's favorite toy in your hand in front of him, have a fantastic reward available in the other hand to give your dog in exchange for it.

Bring the treat to him. Hold the other end of the toy and say "give" right away. When he delivers it into your hands, say "yes" and give him a small treat as a reward.

To train him to drop the toy, do not hold the other end and say "drop" as you show the treat you will give him in return for the toy. When he drops the toy on the ground, praise and reward him.

Teaching both commands is helpful. You never know when your dog might be chewing on something, and you need him to spit it out.

Most of our training puppies are chew-obsessed Labrador and Golden Retriever breeds. The sturdy KONG Classic is one of our favorite toysfor the "give" or "drop it" commands.

13. “Place” Command

The command "Place" is really helpful. You can direct your dog to a certain place, such as a dog bed.

To ensure that the dog knows what you want, leaving the bed in the same place is crucial.

You should also teach your dog to stay down so that he can chill on his bed as you and your friends snack on chips.

The Place Command: How Do I Teach it?

Leash your dog, take him to the bed, and give him praise and treats when he stands on it.

If he does not walk to the bed automatically, putting a reward in front of his nose to cause him to pursue it, luring him onto the bed. Praise and reward him once all four feet are on the mattress.

As you lure him, add a "place" command, praise "good place," and reward him with a treat once he is on the bed.

Have him do a down/stay there once he understands what place means.

If you are persistent, he will eventually be able to get to the "place" by himself.

14. “Spin” Command

Some dogs will instinctively circle and chase their tails. In this trick, you train your dog to go round in a circle.

This party trick is fun. Additionally, it might help a dog gain confidence and use some of their extra energy.

The Spin Command: How Do I Teach It?

Place a goodie in front of your dog's nose. CAREFULLY spin his body in a circle. You'll need to draw a bigger circle for a lab than for a Shih Tzu. Say "spin" as you lure. Once a spin circle is completed, give a small goodie and say, "Good spin."

Always remember to lure either clockwise or counterclockwise in the same direction.

You can teach the other way by using a different cue, such as the twirl.

15. "Catch It" Command

This trick is fun. Your dog might become the next Frisbee champion! Not all dogs are inherently good catchers.

Using a soft toy that your dog loves is, in my opinion, the best approach to teach this.

Most dogs enjoy soft, squeaky toys, but having a range of toys is crucial as even dogs' favorite toys can get boring at times.

How Do I Teach the "Catch It" Command?

Throw the toy gently over your dog's head while commanding "catch."

Before throwing the toy, be sure your dog is paying attention.

Make a scene when he catches it.

Once he understands the notion of catching something, you can have him catch other items, such as a small treat.

16. "Speak" & "Quiet" Commands

Many dogs like barking. When someone knocks on the door, they raise the alarm. They are speaking with us. However, excessive barking is irritating.

Thus, training a dog to bark only when necessary is helpful.

Shelties, a breed known for its loud voice, are one of the breeds I own. When they herd, they let out a high-pitched wail.

It was essential that I trained them only to speak when spoken to prevent me from going deaf and to appease the neighborhood.

How Do I Teach the Speak/Quiet Commands?

Once your dog barks, reward it by praising it and giving it a small treat. You can try to get him to bark by ringing the doorbell or pounding on the door if he doesn't already.

Immediately he stops barking; tell him to be "quiet," and acknowledge and reward the silence.

ADVICE: Once you've practiced these commands, try praising the silent command more than the talk command.

Give your dog something to do after the brief training session, such as a stuffed Kong, to stop him from barking.

Always make him "work" for such treats by following commands first. I suggest the black Kongs for strong chewers. There are softer ones for puppies and seniors, as well as various strengths for those who chew regularly.

17. “Count” Command

You can ask your dog to "count" when he can talk and remain silent when commanded. This clever party trick will dazzle your friends.

How Do I Teach the Counting Command?

Request that he count to five. After five barks, use your "quiet" cue and give him your "speak" cue. To make it less evident, You can combine the "speak" cue with a hand signal, like pointing at your mouth.

Then you can add a hand gesture, like placing a finger in front of yourmouth, to tell your dog to cease barking.

You can slowly fade the verbal cues once the pup has mastered the hand gestures to make the audience more pleased with your dog's mathematical abilities.

18. “High Five” and/or “Shake” Commands

Everybody is impressed when a dog high-fives on command. Introducing this is preferable once your dog has mastered the command to sit.

How Do I Teach the High Five/Shake Hands Command?

You can take your dog's paw and gently pull it off the ground if he is a good handler. Say "high-five" or "shake hands" and give the treat. Repeat this three times.

Use your command cue, reward him when he lifts it on his own, and praise him.

Another approach is to provide the command signal while lightly tapping behind his elbow. When he raises his paw, give him praise and rewards.

Your dog should eventually lift his paw on demand after regular training.

19. “Wave” Command

You're being visited by pals. You can have your dog wave at them as they leave. They won't want to go because they will be so amazed.

You can teach your dog to wave once he has mastered the handshake or high-five.

How Can I Teach the Wave Command?

Tell your dog to give you a high five on cue. The first several times, catch the paw like you typically would. Then, give the cue without touching the paw.

Once he learns the notion, he starts introducing a new verbal signal, such as "wave." You can even add a hand gesture if you wave at the dog while you say it.

You will eventually be able to wave, and your dog will reciprocate! Everyone was delighted when I waved at my dog, who I had adopted from a shelter, and he waved back.

20. “Crawl” Command

This maneuver can steal the show. With any trick, what you say is often what gives the illusion of polish. You can use a phrase like "crawl like a bug."

You can train your dog to crawl once he understands the down command well.

How Can I Teach the Crawl Command?

Put your dog down and hold an enticing treat directly before his nose.

Move the treat an inch away from his nose along the floor. Command him to crawl.

When he advances a little, praise him and give him the treat as a reward.

By utilizing the reward to entice him forward, you can gradually increase the distance and get him to crawl.

Start offering the verbal cue alone without the treat bait once he has crawled a few inches. Keep luring him with the treat up until the point that he truly comprehends what crawl indicates.

He will eventually learn to crawl without the treat bait if you are consistent in your training.

21. “Touch” Command

Using this command, you can train your dog to touch anything with either his nose or his foot. It is known as targeting. Teaching dogs to touch your hand is useful.

It can train a dog to focus on you. It can educate a puppy to treat peoplegently.

How Can I Teach the Touch Command?

Put your hand about an inch or so in front of your dog's nose while he is sitting. Your fingers should be pointed to the floor, and your palm should be towards his nose.

Say "touch" and compliment your dog ("good touch") when his nose softly contacts your palm. You can also reward him with a small treat.

Tips For Teaching Dog Training Commands

Train in quick bursts of no more than 15 minutes each, three times each day.

Do not do too many repetitions of each command.

Finish on a happy note.

Start sessions without distraction and introduce distractions when the dog has mastered the command.

Use a cheerful voice to teach them what "good" means.

Exercise your dog beforehand to prepare your dog for training.

Use rewards and praise as positive reinforcement, which your dog loves.

Ensure your dog is famished, and avoid training him shortly after ameal.

Be consistent in the commands and methods you use.

Timing is crucial. Ensure you compliment and encourage the dog when it exhibits the desired behavior.

When you note the desired behavior, have your incentive goodie available.

Your dog must perform a command before receiving a treat or chewtoy.

Essential

Treats: I've been training my dogs with treats for at least 15 years. Zuke's Mini Naturals and Crazy Dog Train-Me Treats are currently my favorites.

If you have a forgetful memory like mine and a dog like Stetson, you might want to use a treat pouch instead of your pockets because your dog will gnaw through your pockets to get to the goodies you accidentally left in your trousers. The PetSafe brand treat pouch is my favorite.

Still Important

Dog toys are still important, especially if you have a destructive dog likemine. My favorite is KONG Classic.

Use a mat or blanket similar to the one in your bathroom as a dog bed.

Clicker – I trained Archer, Raven, and Elsa with the clicker. Any clicker would work; however, the Mighty Paw Clicker is the one we recommend because it has a fancy attachment.

Harness – Easy Walk No Pull Dog Harness is our preferred harness. Raven, Archer, Linus, Elsa, and Stetson have all worn the Easy Walk Harness.

While teaching all of these commands is vital, the most crucial thing is to exercise your dog. Don't forget that mental and physical exercise is necessary for your dog.

CHAPTER 8: HOW TO TEACH YOUR DOG AGILITY

If you've ever watched dog agility on television, you know that it's an exciting task for both dogs and their owners. However, putting it into practice with your dog is far more enjoyable than just watching!

Agility offers advantages beyond physical fitness; it's excellent for mental stimulation (especially for high-energy dogs) and fostering the relationship between owner and dog.

What is dog agility?

During dog agility, a dog and its handler must navigate a timed, fast-paced obstacle course. The dog must complete some obstacles, including tunnels, jumps, weave poles, and ramps, in a predefined order.

Advantages Of Teaching Your Dog Agility

The benefits of teaching your dog agility are endless!

Great exercise: Running, climbing, weaving, jumping, all at a fast pace, will undoubtedly exhaust them. Agility is especially ideal for active, high-energy dogs who require lots of exercise.

Reduces boredom: Agility gives your dog cerebral stimulation in addition to physical activity, which lessens boredom and the destructive behaviors that might result from it.

Encourages positive behavior: Dogs rely on their owners to give them the orders they require to finish an agility course successfully. Dog-taught agility will give you more attention and be more obedient to your directions.

Increases the tie between owner and dog: Many agility handlers mention how teamwork has strengthened their relationship with their dog.

Dogs with strong activity levels and a lot of exercise benefit greatly from agility. Additionally, it stimulates the mind, which lessens boredom and the potentially dangerous behaviors that might accompany it.

Is my dog suitable for agility?

All adult dogs in good condition can participate in agility; however, extremely older or ill dogs with health problems should not. It's best to start gently with a puppy and wait until they are between 12 and 18 months old to finish the entire course. For dogs that have joint problems, such as arthritis, or those who are prone to back injuries, such as Basset Hounds and Dachshunds, you might also need to modify or omit specific parts, such as jumps.

Many pet owners believe agility is especially beneficial for dogs who have common behavioral problems, such as:

Anxious or Nervous dogs: Many nervous or anxious dogs find that ambiguity stresses them out because they never know what will come next. Agility training can benefit these pets like daily routines and repeated tasks. They pick up the sport easily and love the repetitive patterns, which help to control their anxious or tense behavior.

High-energy working dogs: If your dog is a working breed, you may find it difficult to provide enough activity to burn off all of its energy. Agility is often a physical and mental challenge for these high-strung dog breeds, which can also help exhaust them.

Dogs who struggle to focus: Agility courses can be a good option for your dog if they continually fidget or wriggle and struggle to concentrate, either at home or in a training class. These courses will help them channel their energy and improve their ability to concentrate.

Is agility suitable for me?

It's important to consider if dog agility is an appropriate exercise because it should be enjoyable for you and your dog. Suppose you don't already have a strong basis for fitness. In that case, you should develop a regimen of training activities that will help you gain strength, agility, and balance so you can support your dog as they navigate the agility course. Agility can be a fast-paced activity.

Ensure you go slowly until you feel at ease finishing a course with your dog.

What To Know Before Introducing Agility To Your Dog

Establish a base of basic obedience.

The agility course will need your dog to closely follow your instructions, so refresh her fundamental obedience skills with positive reinforcement methods, such as sit, stay, and come. Mastering stay is crucial when using the teeterboard, where your dog must remain motionless until the far end reaches the ground.

Increase the focus of your dog.

Make sure your dog can remain focused on you despite distractions. There might be a lot of distractions while you're on the course, including other dogs, loud noises, and of course, all the entertaining obstacles to play with! You can aid with this by practicing commands like "Look at me" or "Watch me" while you are out and about.

Get your dog used to various movements.

Before exposing your dog to obstacles, get her used to moving in unusual ways. Teach your dog to tread on top of objects, crawl through, and climb over them. You can even teach her to walk backward and put her paws in certain locations. Throwing treats in that direction will teach her to tightly turn around an item and to go away from you or to the left and right.

Teaching your dog agility at home

Build your agility obstacles at home to see if your dog loves it. Here are some pointers on how to build your obstacles and teach your dog the basics.

Jumps

Ask your veterinarian beforehand if jumping is safe for your dog. Jumping is not advised for some breeds because they may be more prone to back issues, such as Basset Hounds and Dachshunds.

Once your veterinarian has given the all-clear, you can create jumps by placing a piece of plywood on top of a stack of books. Ensure that if your dog knocks the jump over, she won't injure herself. Start low and gradually raise your jump height. Start one to two inches above the ground for large breeds; you might want to start with the board on the ground for little breeds.

Tire Jump

An old bicycle tire or hula hoop ring can be used for the tire jump. Just make sure your dog can get through it without difficulty. Hold your tire firmly as your dog first walks through it. You can gradually lift it till you can hang it from a tree branch.

Dog Walk

The dog walk is an elevated walkway with ramps at either end. You can use a picnic seat with wooden planks at either end to serve as ramps.

You can try starting with the end part first because some dogs could be reluctant to climb up onto this obstacle. Pick up your dog, then set them down close to the obstacle's end. Dogs will typically take the few steps necessary to get off. Then, increase the difficulty level until your dog is content with the obstacle.

Tunnel

A plastic children's tunnel can be cheaply purchased at a department shop, or you can create a fake tunnel by draping some sheets over several chairs. Don't push your dog inside the tunnel if they seem hesitant. Try enticing them with treats or sticking your head through the opening and calling them from the other side.

Weave Poles

The weave poles, one of the most fascinating obstacles to watch, typically comprise 10 to 15 vertical poles your dog must navigate. You could use PVC pipes buried in the ground or ski poles.

Start with the poles pretty far apart. To acclimate your dog to the weaving motion, stroll around the poles while they are attached to a leash. Next, let your dog walk between the poles on her own. As your dog starts to master the moves, you can gradually bring the poles closer together. Flexibility is crucial in this situation, so go slowly to prevent your dog from hurting themselves.

Teeter Board

One of the most challenging obstacles for many dogs is the teeterboard, which demands a lot of confidence around moving items.

Start with low items and train your dog to get used to things moving underneath them, such as a toy wagon, a wobble board, or a skateboard. Reward your dog whenever they approach the object, touch it with a paw, and finally, when they balance on it. You want to make it an enjoyable game and foster a favorable association with moving items.

When you think your dog is ready, you can make your own teeter board by stacking a long piece of wood on top of a pipe. Once more, you must proceed with caution while enticing your dog with goodies and rewarding them anytime they touch or take a few tentative steps onto it. Once they reach the middle and the teeter shifts, give your dog a jackpot of goodies for staying on the teeter. You can switch to thicker pipes as soon as they are at ease with the movement.

Dog agility is a frantic, thrilling challenge that is fun for dogs and people. In addition to being entertaining, agility training can be a terrific method to assist your dog in learning obedience, increase their attention span, and help with behavioral problems like anxiety by offering them an opportunity to expend their energy.

If you and your dog enjoy agility, you can start with simple agility at home by building jumps, tunnels, and teeter boards or try your hand in a more structured dog agility class.

CHAPTER 9: SEARCH AND SNIFFING GAMES

Dogs like sniffing, and their noses are strong tools. Dogs "see" the world through smell, with over 220 million scent receptors in their noses, over fifty times as many as ours. It should be no surprise that dogs can serve as scent hunters, finding lost individuals, truffles, explosives, narcotics, and even chemical changes that signify low blood sugar, cancer, or other human ailments. What a dog's nose can detect is incredible.

Although pups are blind at birth, they have a fully functional and active sense of smell. The most powerful sense in a dog is the one we humans tend to ignore the most. While humans concentrate on how things seem, our dogs are more interested in the multitude of aromas floating through the air. They are interested in what the neighbors had for dinner, which raccoon passed through the yard the night before, and whether or not the retriever across the street recently took a bath.

For dogs, sniffing is a normal and often relaxing behavior. Dogs are mentally stimulated by sniffing various things and absorbing information. Your dog's life is greatly enriched by giving it lots of smelling time. Imagine how much mental workout your dog receives when they learn to locate a specific scent on cue, considering how much of their brain is devoted to evaluating smells.

Scent training is a fantastic method to strengthen your bond with your dog and enhance their mental well-being. It's a great way to teach your dog new abilities, have fun together, exercise them, and mentally challenge them. And to begin this fascinating pastime, you don't have to leave your home.

Dog Scent Training Supplies

You'll need a few items for your dog to learn scent training. Some may already be in your house, but you can find them all online.

Scent source. Fragrant oils are a useful tool for scent training. Anise, clove, and birch are some of the most widely used scents. You can start with truffle oil if you already have a certain sort of scent work in mind for your dog, such as truffle hunting. It's crucial to remember that many essential oils can irritate and hurt dogs, so make sure the oil you pick is safe for your dog, and only use them in very small amounts when using them expressly for scent training.

Cotton swabs. These are used for carrying scented oil.

Disposable gloves. It's crucial to avoid mixing your scent with the training scent.

Plastic sealable bags. This prevents any leftover ingredients from contaminating the smell you're practicing with.

A scent tin. This has to be a little container accommodating the cotton swabs, such as a tin or jar. The top of the lid should have a few little holes drilled into it. A well-cleaned and dried spice jar or saltshaker will work.

A scent box. Utilize a plastic container with drilled holes in the lid. You will use it when you begin training your dog to detect scents.

A Prize/Reward. Your dog's preferred expensive treats.

Most domestic dogs nowadays are a little out of practice using their noses, which they once relied on to live. The good news is that you can help your dog use this latent sense of smell by adding enjoyable scent games to her repertoire of behaviors with only a little planning and patience.

How to Sniff Train Your Dog

Make a fragrance product. It's crucial to apply the fragrance to one item; contaminating your hands, clothing, or other items with the same scent might cause your dog a lot of confusion. You will thus require gloves and plastic bags.

Do this in a room far from your dog, such as in a restroom or outside the garage with the door locked. Put on your gloves, apply two drops of the oil on a cotton swab, and then put the swab in the scent tin with the lid screwed on. Put your gloves in a plastic bag and seal them before throwing them out with the trash.

To Introduce the scent to your dog. Hold out your fragrance container for your dog to sniff. Reward them right away if they give it a sniff. Give your dog the reward right where the smell is coming from. Reward them with a treat each time they place their nose on the smell tin. Continue giving them treats if their nose remains pressed on the tin.

Switch which hand is holding the scent tin while you continue doing this for a few minutes. Avoid going too hard during each training session. You want your dog to continue paying attention and create a connection between the smell and a tasty reward. If you put in too much time, they can grow bored, or their noses might become overwhelmed by the new scent.

Teaching your dog smell detection. Once your dog can dependably locate the scent tin, it's time to increase the difficulty. You should also provide a verbal signal to instruct your dog to look for the scent at this stage. Simply begin by saying the word you want to use, such as "find," "seek," or "search."

Put the scent tin inside the plastic scent box. When your dog detects the smell in the box, practice praising her by giving her treat there. Start by setting the box down next to your feet, then move it to either side of you.

Make it a little bit more challenging each time you play this game. While your dog is present, you can "hide" the smell box. Every time she sniffs out the smell box, throw a huge party!

You may raise the difficulty level as your dog becomes more adept at tracking the smell. Start concealing many fragrance boxes, just one of which should have the scent tin inside. Alternately, ask a friend or member of your family to disguise the smell so you won't unintentionally offer your dog any cues through your body language while they search for it.

Scent Training Games for Dogs

It's not necessary to teach your dog a brand-new smell to enjoy nose work. Instead, you should push her to use her nose to sniff for their preferred treats or friends. Here are a few entertaining nose work games you can play at home with your dog.

Exciting Scent Games Your Dog Will Enjoy

1. Find the Food

This technique is easy to activate your dog's keen sense of smell. You only need to scatter goodies all through the house, hoping she will find them by smelling them. She will immediately become aware of the possibility of locating other ones with her nose after finding the first one (often accidentally). When she is outside the room, start by putting out one or two goodies. then give her a call. She'll enthusiastically eat them up and look for more. It is advisable to repeat this process, but start hiding the treats this time. Consider placing them in a corner, straight behind a sofa or coffee table, or even partly beneath a dog cushion. Put them down when she is outside or somewhere else in the house. Allow her to go and locate them. You'll soon notice her smelling for them.

Vary quantity and placement; some days, only hide one treat. Whenever she finally "gets it," switch out the secret item. Try hiding a food dispenser toy with treats inside. Hide a feather covered in cheese. Place a frozen broth or broth cube (on a dish) in plain sight. Using the same technique, gradually move it outside into the yard, making it easier before making it harder. Try hiding a chicken egg out there! You can even attempt this in your automobile or the home of a friend.

2. Pick the Hand

Here's a fast tip to get your dog's nose working. First, gather some mouthwatering bite-sized treats that you can hold in your palm. Due of kibble's weaker fragrance, a little piece of cheese or turkey meat will work better. Next, place one in your palm and create a loose, palm-down fist. Then, while your dog sit in front of you, hold out your hand for her to check it out. While doing so, say, "Find it!" When she's given it a sniff, say, "Good, find it!" while holding out your hand with the reward in it. Repeat this a few times. Add a second fist that is empty. Avoid letting her see which hand you placed the treat in. Then, while keeping your hands closed, move them back and forth. Finally, offer both to her and say, "Find it!" When she sniffs, say, "Good, find it!" and hold out your treat-holding hand. Switch which hand is holding the goodie as you go through this procedure. Before opening up and moving ahead, wait until you can see her nose extremely "alert" on the reward hand. The idea is to show to her that a treat's position might vary and that smelling it out is the only reliable method of locating it. Once she gets it, make it more difficult for her by adding a friend's two fists.

3. New Animal Scent

Dogs are inherently hunters, rivals, and trackers of prey. Take advantage of this by releasing the fragrance of a new animal into your dog's yard to see whether she can smell it. Try it just outdoors as dogs often mark their territory with animal urine.

Ask a friend to give her dog or cat a thorough rubdown using an old towel or cloth. They should try to get a drop of urine on the fabric because it stinks so bad. If not, rubbing it will do. When your dog isn't around, hide the cloth in a secluded spot in the yard, such behind a bush or a tree. After that, let your dog out to see what happens! Use a variety of animal odors and do this at random to keep your dog on their toes. After experimenting with cat and dog scents, try parrot, hamster, ferret, and any other scents you can locate.

4. Hide and Seek

Here's one where you are the treat. When your dog is distracted somewhere else in the home, hide in a closet, behind a bed, or somewhere else she wouldn't normally expect you to be. Then just wait. She'll definitely start looking for you. When she finds you, praise and reward her.

Then, take it outdoors, ideally by yourself, to a wooded area where dogs are permitted to run free. As you leave the area and go for shelter in the woods, ask a friend to watch your dog. After waiting for thirty seconds, your companion should call out, "Where's (your name)!" and then let her go. Your dog should leave, her nose to the ground searching for you. You should lavishly compensate her when she momentarily finds you. Increase your range gradually so that she can locate you wherever you are.

5. Shell Game

Obtain four sturdy, coffee-cup-sized containers that she will have a hard time knocking off or shattering. Glass and paper should not be used as they are both delicate and prone to breaking. While your dog sits and watches, place a treat below one cup and move it back and forth. Then, command, "Find it!" Lift the cup when she sniffs it and say, "Good, find it!" while she consumes the treat. If she tips the cup, that's okay. Add one more cup. Place the treats in the cups and rock them back and forth gently. Say, "Find it!" and let her sniff the cups individually. Wait till she sniffs the right one before praising her and elevating the cup. Continue doing this until she constantly picks the right cup. Once she constantly succeeds on the first attempt, add a third cup and keep going. You can see at that moment that she is using her nose and not simply picking it at random.

6. Where's Dinner?

Unlike wild dogs, who must hunt and locate food every day, our dogs know they will find a meal in the same place every day. But what if one day you called her to dinner and found her dish somewhere else? Simple: she would start looking for it right away. Put it in the room next door first; she will begin smelling there right away and find the fragrant dish in no time. The next day, stow the dish away in your home and ask her around for supper. She will ultimately discover it, though it could take her a bit longer, and she will consume it. Once she gets the hang of this activity, move her bowl once or twice a week and ask her to search for it.

7. Scent Trails

Dogs like smelling items outside food as well. Dogs' natural tracking abilities can be stimulated by the unique scents of essential oils like valerian, anise, and lavender. Apply a few drops of essential oil on a toy (a ball works nicely) to get things started. After that, play a quick game of indoor fetch before giving a reward. Do this many times per day.

The next day, hide the same toy when the dog is away, then scatter a trail of bread crumbs-like pieces of oil-coated paper 20 feet away from the ball. Allow the dog into the room where the trail begins and tell it to "Find your ball!" Most dogs will first smell the paper before smelling the ball. Carry on and commend her when she follows the path. If necessary, show her the first perfumed paper to get her going. When she locates the ball, give her something to celebrate! Reduce the amount of scented sheets gradually until she can find the scented ball on her own. Once it has been mastered indoors, move it outside into the yard. Change the aroma and the toy, then start again. Anything your dog likes, such as peanut butter, cream cheese, and chicken fat, can be used.

8. Find the Scent Itself

You can teach your dog to look for a scent itself, instead of using it as a means to locate a ball. This is a simplified explanation of what bomb- and drug-sniffing dogs do.

If you've taught her to find a ball by following a scent trail, your dog has already acquired the ability to concentrate on smell. Place the same scented ball first in a shoebox. Then tell her to "Find your ball!" so that she will go up to it and take a sniff. When she starts to claw and scrape at the box, you should take away the ball (if she hasn't already) and praise her. Keep the ball in its original container and repeat the operation using just three boxes (to prevent cross-contamination). Tell her to keep trying and to "Find your ball!" until she does. Reward her with a fast fetch session.

Now, instead of putting the scented ball in one of the three boxes, place a slip of paper in the same box with a few drops of the same essential oil on it. Hide the scented ball outside, wash your hands, then put a fresh, unscented ball in your back pocket. Say, "Find your ball!" again, encouraging her. Praise her loudly when she settles on the scented paper box, then grab the ball from your pocket and toss it for her as a treat. Increase the distance and quantity of boxes as you go on. She'll quickly gain experience as a tracker!

These simple smelling exercises are far from everything that a dog is capable of in terms of tracking.

CHAPTER 10: DOG GAMES

There is no shortage of dog games to play! Finding the ideal activities for you and your dog is crucial, as not all games are equal. To help narrow your search, we have divided the entertaining games to play with your dog into categories, including DIY games, games to keep dogs smart, indoor games, sense of smell games, games to keep dogs fit, and activities for older dogs.

DIY GAMES TO PLAY WITH DOG

Don't want to spend much money, but you need a new game? You can use objects you already own from around the house to play the games below!

1. Muffin Tin Puzzle

Tennis balls, snacks, and a muffin tin are all required for this game. Put some goodies in each muffin cup, cover them with tennis balls, and then give your dog the tin to play with! Once your woof learns how to get the treats, you can increase the difficulty by placing treats in just a few cups while covering them with tennis balls. Your dog must use their sniffer and problem-solving skills in this game!

2. Treat Burrito

You only need a towel and some little treats to make a treat burrito! Spread the towel flat, then disperse treats over the surface. Slowly roll the towel up, keeping each treat in place. Serve the burrito to your dog. She will have a great time unrolling the towel to get each treat.

3. Cardboard Tube Treats

Paper towels and toilet paper rolls made of cardboard can be fashioned into dog toys! One game option is giving your woof a cardboard tube filled with peanut butter to lick up. You can freeze the tube if you want to keep Fido busy longer!

Another great game to play with your dog with cardboard tubes is to place numerous tubes in a shoebox so that they all support one another and stand straight up. (All of the tubes need to be about the same height.) Put treats or kibble into the tubes, then present the puzzle to your dog! Your dog will enjoy hunting for every bite.

4. Snack Track Down

Any empty container, even yogurt cups, and shoeboxes (after being well cleaned), can be used for this game. If you have a large dog, you will need a larger container, and vice versa.

The game should be set up when your dog isn't present. Put a few containers on the ground, at least one of which should contain a smelly treat. Bring your dog in to sniff and find the treat!

You may want to drill holes in the containers to make the treats easier to smell when you first play this game. As your dog gets better at the game, you can increase the difficulty level by hiding treats in only one container, with fewer containers spaced farther.

5. Ball Pit-Palooza

You will need an empty kiddie pool and lots of balls to properly excite your dog's brain and nose! Fill the pool with balls, then sprinkle treats on top. The balls and treats will move when your dog enters the pool to grab the treats! While it's one of the more difficult dog games, dogs find it incredibly gratifying.

This game and other dog games listed above are excellent choices for feeding a fast eater. If you disperse a whole serving of dry food in the swimming pool, your dog won't be able to consume it as quickly as usual.

GAMES TO MAKE DOGS SMARTER

Playing with dogs can be both entertaining and educational. The dog games listed below can boost your dog's IQ.

1. Hide-and-Seek

Your dog can play hide-and-seek with you if you ask them to sit still while you look for a hiding location. As soon as you're hidden, tell them to "come" locate you. Treats should be given to him for his successful quest.

2. Treasure Hunt

Similar to hide-and-seek, a treasure hunt involves hiding a smelly treat (or toy) rather than yourself. Then hide the prize after telling your dog to sit and stay. Once it's hidden, give your dog the all-clear by saying "okay" or a similar command, and the hunt begins!

3. Stop and Go

Playing some fun games with your dog might help to reinforce training. Dogs can practice their "come" and "stay" commands using the Stop and Go game.

Get your dog fired up and enthusiastic, then create some distance between you and her. Command your dog to "come," but before they can approach you, tell them to "stay." Give your dog a treat if they manage to stop in their tracks. Play on by performing this action several times! Stop and Go helps your dog learn self-control and improves his listening abilities.

5. Name Game

According to the American Psychological Association, the typical dog can learn 165 words! Your dog probably doesn't know that many words yet. Using the names of their preferred objects, people, and activities is a fantastic approach to teach them new words.

I believe naming a toy is the easiest. If your dog loves a particular stuffed rabbit, name it "Bunny" and use it whenever it is playing with it. You should also use it whenever you give the toy to your dog or take it away. You can test your dog's skills by setting out Bunny and two more toys, then instructing your dog to "Get Bunny."

DOG GAMES TO PLAY INDOORS

You can have fun with your dog without going outside! Here are some fantastic dog games for late nights, rainy days, and people who don't have yards!

1. Clean Up, Clean Up

Looking for entertaining dog games that are also good for you? The winner is clean up! Playing this game, you can practically educate your dog to clean up after themselves. Teach your dog to throw their toys into a laundry basket or bin. Use orders like "retrieve," "drop it," and "put it away" to teach your dog how to play this game.

2. Snuffle Mat

If you want to keep your dog busy for a while, it's time to pull out the snuffle mat. Your dog must use his nose to find treats that are concealed in folds and pockets on snuffle mats. This game is ideal for playing with dogs indoors because it is quiet and clean.

3. Obstacle Course

A home obstacle course is a great way to test your dog's agility. Clear up some space, then create your course with home objects. Your dog might jump over a hula hoop, wriggle under tables, bounce over blankets, and weave past stacks of books! Obstacle courses provide excellent mental and physical activity even indoors.

4. Staircase Run

If your home or apartment has stairs, you can play engaging and active dog games there. Playing the game of "racing" your dog up the stairs will provide you both with some helpful exercise. But be careful not to run too quickly because you can fall!

You can play a retrieve version using the stairs if you don't want to run. To train your dog to chase a toy, instruct your dog to "stay" at the bottom of the stairs while you throw a toy up to the top. Your dog will be pooped out after a few runs!

Important Reminder: Older dogs, growing puppies, and dogs with joint problems or arthritis shouldn't play these stair games. Additionally, carpeted stairs are preferable to wooden or concrete surfaces.

5. Obedience Training

Start training in a quiet, distraction-free setting, such as inside your home. Show your dog a treat or a toy first, then praise them as they approach and reward you. After a few practice sessions, anytime your dog looks at you and begins to approach you, add your selected verbal signal (come, here, etc.). Only use the cue when you are certain that yourdog is approaching you.

You can gradually increase the difficulty by asking your dog to come before revealing the treat. But when they get to you, reward them with a high-value delicacy like chicken, cheese, or beef liver. Additionally, try extending the distance within your low-distraction environment.

With some treats and positive reinforcement, learning can be enjoyable and regarded as a game! Also, teaching your dog instructions like "come" will make their games safer. Dogs with sound training obey their owners and remain disciplined, even when having fun!

GAMES TO KEEP DOGS FIT

Many dog games double as exercise! Exercise your dogs with the games below.

1. Tug-of-War

While tug-of-war is a well-known pastime, did you realize that your dog can also benefit from the pulling and tugging? You and your dog can play tug-of-war outside or inside; just a little room is needed. You should challenge your dog while also letting her win occasionally! (Winning every time is not fun!)

Important Note: Stop playing tug-of-war with your dog as soon as its teeth approach your hand. This will teach your dog that biting you is inappropriate and that the pleasure will end immediately. It's important to teach your dog to drop the rope when commanded. They need to be aware of your dominance as tug-of-war is a game of domination to them.

2. Fetch

Another classic that burns a lot of calories is fetch.

Use a ball or your dog's preferred toy to play a typical game of fetch with your dog. Throw it as far as you can, then do several burpees, lunges, situps, and/or pushups while your dog fetches and returns the toy. Continue until both of you are exhausted!

You can use a tennis ball or other ball designed for retrieval to playfetch. Play fetch with a frisbee if you want to truly thrill your dog!

3. Sprinkler Fun

Sprinklers provide entertaining activities that you can play with your dog and kids! If you have young children and furry family members, playing in the sprinkler is a terrific way to entertain the whole family on a hot day. Install your sprinkler when the grass is dry, then show your dog how to leap over the water. This is a fantastic technique to raise a dog's heart rate and keep him cool at the same time.

Important Note: While you and the kids should keep drinking water from inside the home, your dog can enjoy a taste of sprinkler water.

4. Soccer

Playing a dog-friendly version of soccer is possible! Get a soccer ball and train your pooch to nudge it with his snout. Offer treat for his "passes"! Finally, try gently kicking the ball toward your dog to see if he can return the favor. Give your dog a chance to play defense if he excels at passing! Dribble the ball up to him, and see if he tries taking the ball away.

Your dog will become a regular football player before you know it!

5. Chase Bubbles

How many bubbles can your dog pop? Blow a few bubbles at a time in front of Fido. Point to the bubbles as you pop a few to show your pup how it's done. Your pet will soon be playing with bubbles and romping around! Remember to wipe the soapy suds off your dog's face when you've finished playing.

SENSE OF SMELL GAMES FOR DOGS

Dogs have very keen noses, as the scent is their most acute sense. Thus, playing with dogs is more enjoyable when there are odors involved!

1. Which Hand Game

You've undoubtedly participated in guessing games like this one yourself, such as when someone holds up their two fists and asks, "Which hand has a quarter in it?" This isn't at all a guessing game for adog, though. Everything is dependent on smell!

When your dog isn't looking, sneak a treat into one of your hands to start the game. Present your dog with both hands after making loose fists out of them. While your dog might first sniff a little, they should eventually focus on the hand containing the treat you will give them as a reward.

Invite a buddy to play to increase the difficulty and expand the game's possibilities! Who does not like playing with dogs?

2. Where's My Dinner?

Making your dog work for his food is the object of this game! Even though most dogs' meals are delivered in the same location every day, they find it quite enlightening to search for their dinner. Distract your dog or put him away to play this game before preparing his dinner. The meal should then be served in a location close to but distinct from his regular dining spot.

The dog will begin searching for the meal after you release him (or call him for dinner), and he notices that it isn't where it usually is. Your dog utilizes his instincts while also getting a wonderful mental workout! He'll discover meals quite fast if his nose abilities are good. Changing the meal location farther from your dog's initial setting can make this game more challenging!

3. Stroll and Sniff

Walking is an excellent way to let your dog use his nose, as there are many scents to discover when you're out and about. You can transform a walk into a game by taking some small treats and sometimes stopping to drop one into a grassy area. Before you throw the treat, show it to your dog, and then allow them some time to find it.

Important Note: Avoid playing this game on another person'sflowerbed!

4. Treat Trail

Lay down a brief trail of small (low-calorie) treats with a larger treat waiting at the end while your dog isn't nearby. Once you've shown your dog where the trail begins, watch them follow it to their treat! Once your dog has mastered short trails, you can make it longer with more spacing between each treat.

5. Find that One

Playing this game with your dog will make you believe she is a genius. Bring several dog toys into an open space, either outdoors or indoors. Spread them out on the ground. Then, get your dog and start playing with one of these toys. You should make a lot of contact with the toy to ensure that your scents stick around.

Once you've played with the toy for a while, hold your dog by the collar and throw the toy back among the others on the ground. Tell your dog to "find that one" and let go! The objective is for the dog to bring to you the toy you both were playing with. It's not just about picking out the newest toy in this game but also the one that most strongly smells like you and your dog!

If the dog brings the wrong toy back, take it and request that they "find that one." Once she brings you the right toy, it's treat time!

GAMES FOR OLDER DOGS

Despite popular opinion, an aging dog can learn new tricks and games! The dog games described below are excellent choices for dogs with joint issues, arthritis, or other health issues, and they are also suitable for older dogs. Some of these activities can be treated like games while they are truly physical therapy.

1. Swimming

Swimming is one of the best games to play with your dog if they enjoy the water, whether young or elderly. Swimming is a low-impact activity that is gentle on the joints and muscles but a lot of fun for your dog. Make sure your dog can doggie paddle before you take them swimming; if not, you might get them a life jacket.

2. Cavaletti Course

This game is a scaled-down obstacle course that is ideal for older dogs. Place a few poles, such as broomsticks, about a foot apart to set it up. After that, softly lead your dog around the course while keeping a short leash on them. This is a great strength-training exercise, and you can make it harder by raising the sticks one or two inches using pillows or crushed soda cans.

3. Modified Fetch

Fetch is a classic dog game to play! Your dog can still play a modified form of fetch even if they aren't as mobile as they previously were. Instead of throwing the ball, you can think about rolling it on the ground. Instead of a sphere, you might use an octagon ball as it won't roll as quickly or far.

Important Note: Retrieving distances for dogs with aches and pains should be kept short, and you should be mindful of your dog's energy levels to avoid overexerting them. Understand when to end fetch and transition into nap time!

4. Backward Walking

Hold on, Fido! Backward walking is a fun pastime that strengthens your dog's rear legs and enhances coordination and balance. You can play with your dog by holding up a toy or goodie in front of your dog while telling them "back" as you approach them. Your dog should step back once you are physically close to him, and you'll reward her with a treat. You should make this game brief and simple to prevent straining your dog's limits.

5. Puzzle Dog Games to Play

Puzzle games are the best dog games for all dogs, including seniors! Older dogs can still have extremely bright minds, so it's crucial for you to provide cerebral stimulation for your dogs. The Treat Burrito, Muffin Tin Puzzle, Cardboard Tube Treats, and Snuffle Mat are just a few puzzle games suitable for older dogs.

BONUS CHAPTER

TIPS TO IMPROVE YOUR DOG'S MEMORY

New pet parents might struggle to picture life without their amazing furry buddy. However, do you ever ponder whether your dog feels the same about sharing a life with you? Do dogs have memories? Do dogs even remember their owners?

The extent of dog memory is still debated, but let's look at what is now understood.

Do Dogs Have Memories?

Although dog memories have existed, nothing is known about their exact nature, including how much they remember things.

There are a lot of stories about dogs' memory, but only a few experiments have been conducted.

The good news is that research on dogs' memory is now being conducted, particularly at Duke University's Duke Canine Cognition Center, where scientists examine the following issues: "What cognitive techniques do dogs utilize while navigating or remembering events? Do all dogs travel and retain information similarly? Are there systematic breed differences??" Any of these queries could yield interesting findings.

Dog Memory Types

An excellent follow-up question to a question like "Do dogs remember their owners?" "How do we even know?" because there isn't any scientific data on how a dog's brain "remembers" things. Dogs, fortunately, make excellent test subjects (Sit? Sure! Fetch? You bet!), allowing professionals to deduce knowledge based on a dog's behavioral tendencies.

Dogs can be intelligent, but enough research hasn't been done to say whether or not different dog breeds have different memory capacities.

In general, dogs show a variety of memory cognitive behaviors, such as the ones listed below.

Memory Span

The short-term memory of dogs is quite limited. According to a 2014 research on several species, including rats and bees, "Dogs forget things within 2 minutes," as stated by National Geographic. Unlike dolphins and other animals, dogs appear to have a short-term memory that lasts much longer than those two minutes.

Associative and Episodic Memory

Despite their short memory span, dogs show strength in other forms, such as episodic and associative memory.

Associative memory is the brain's method of establishing a connection between two objects. For instance, getting a cat into their pet carrier could be challenging because they link it with going to the clinic. When a dog notices its owner's leash, she knows it is time for a stroll.

Episodic memory is the recollection of an event that has personally affected you and is related to self-awareness.

It was once believed that only humans and a few other animals possessed episodic memory. However, a ground-breaking study published in Current Biology found solid "evidence for episodic-like memory" in dogs. Previous studies have shown that dogs possess such abilities. The researchers trained the dogs to the point where rather than saying "Lie down," the researcher could just say "Do it," and the dog would comply.

Anecdotally, developing sophisticated cognition in dogs could not be that far off. Dr. Stanley Coren, a renowned psychologist and dog author, reported in an article that he once spoke with a man who had lost his short-term memory as a result of a brain injury during childhood and relied on a "memory assistance dog" to help with episodic "new memories" like where his car is being parked. Interesting thing.

Do Dogs Remember Their Owners?

These results support the hypothesis that adopted dogs may retain memories of their prior owners, though it is unclear exactly how. For instance, a dog that had terrible living conditions would link particular things or places with unfavorable feelings or worried behavior. And it is undeniable that dogs miss their owners when they leave the house. Just see how ecstatic they become when you enter the front door.

This does not, however, imply that your new puppy is pining for another family. Your new puppy will be content to concentrate on the present and enjoy spending time in their forever home as long as you provide a loving and supportive atmosphere.

How To Help Your Dog Improve His Memory

When dogs are being domesticated, we limit their capacity to explore and react to their instincts, which they had previously placed such a heavy reliance on. A wonderful approach to get the most out of a dog and a great method to speed up obedience training is to structure play and training to trigger these instincts and impulses.

Certain breeds, like Gundogs, have a latent desire to retrieve. However, other breeds can also benefit from playing activities that encourage retrieving from memory. The following activity is one that most dogs will like and find intellectually beneficial. Many dogs enjoy learning the hunting retrievers' techniques, even if they never go hunting.

A busy retriever in a hunting situation must recall where multiple birds landed and bring each one back. Hunters refer to this as marking, but anybody can enjoy this difficult game. Start by throwing a few things so your dog can see where it lands. Then, throw the retrieving object away from sight—possibly in tall grass or behind something. It should be easy for your dog to locate and retrieve it.

Then, hold him or have him sit and wait until the object has touched down, then send him to get it. Make him wait a little longer as he gets better, up to 30 seconds. He is now utilizing his memory to locate something he saw land earlier but can't yet see. But can he locate two of these items? In the beginning, probably not, unless he can fit them both in his mouth. He will probably race to one before taking it to the other, where he will either swap stuff or pause and contemplate his options. You must show him how to bring each back to you, one at a time.

If you taught him to fetch using the "hallway trick," you've got a good start. You do it by sitting in the middle of your hallway, throwing a toy one way and then throwing another toy the other way as soon as he gets it back. In this sense, he is already familiar with returning from something and quickly departing for another. However, the hallway is inadequate for large dogs because there isn't enough space. Additionally, you'll discard both objects before he returns one, which makes a significant difference.

One thing is still true: You need to make it hard for him to get the second object without first going past you. One way to do this is to stand in a corner of your home. Another more ambitious approach is to use temporary fencing to divide your property practically in half, then stand at one end of the fence.

In either scenario, you will toss one object to one side of the corner or fence and the other to the other. Allow your dog to see them both land. It will be easier for him if you send him first to retrieve the object you tossed second. Once he has delivered it to you, turn him to face the other object and instruct him to get it. Encourage him to bring it back.

If he appears puzzled, you might need to run some distance with him. You can then move away from the corner or fence as he gains experience, leaving a space between you and it. Call him toward you if he forgets and comes straight to the second item without bringing the first one to you. No one claimed this would be simple, but with time he'll be able to perform this on an open field. And that is when it's really enjoyable.

The goal of the exercise is to encourage the dog to rely on his memory to succeed. The more times you practice this activity, your dog will learn that using memory, even short-term memory, will always end in success and always result in fuss and praise.

CONCLUSION

If you're seeking engaging dog activities, the above methods are a terrific approach to stimulate your dog's brain while still having fun. Exercise is enjoyable for your dog but may also benefit their mental health. Choose a brain training game or activity appropriate for your dog's personality and needs from the many available options. You can quickly observe your dog's behavior change with a little perseverance, work, and yummy training treats.

Aim to give your dog at least 30 minutes of play every day, divided into two periods of 15 minutes each. Dogs require different levels of mental stimulation, so it's crucial to watch your dog and adapt the amount of play to their exact requirements.

Larger, more energetic breeds typically require more mental stimulation than smaller ones.

Don't forget to scan the QR Code to get all bonus content!